FIGHTING IN THE AIR

(Issued by General Staff)

APRIL, 1918

The Naval & Military Press Ltd

Published by the
The Naval & Military Press
in association with the Royal Armouries

Unit 10 Ridgewood Industrial Park,
Uckfield, East Sussex, TN22 5QE
Tel: +44 (0) 1825 749494
Fax: +44 (0) 1825 765701

MILITARY HISTORY AT YOUR FINGERTIPS
www.naval-military-press.com

ONLINE GENEALOGY RESEARCH
www.military-genealogy.com

ONLINE MILITARY CARTOGRAPHY
www.militarymaproom.com

ROYAL
ARMOURIES

The Library & Archives Department at the Royal Armouries Museum, Leeds, specialises in the history and development of armour and weapons from earliest times to the present day. Material relating to the development of artillery and modern fortifications is held at the Royal Armouries Museum, Fort Nelson.

For further information contact:
Royal Armouries Museum, Library, Armouries Drive,
Leeds, West Yorkshire LS10 1LT
Royal Armouries, Library, Fort Nelson, Down End Road, Fareham PO17 6AN

Or visit the Museum's website at
www.armouries.org.uk

FIGHTING IN THE AIR.

GENERAL.

1. THE NECESSITY OF FIGHTING.

The uses of aeroplanes in war in co-operation with other arms are many, but the efficient performance of their missions in every case depends on their ability to gain and maintain a position from which they can see the enemy's dispositions and movements. Cavalry on the ground have to fight and defeat the enemy's cavalry before they can gain information, and in the same way aerial fighting is usually necessary to enable aeroplanes to perform their other duties.

Artillery co-operation, photography, and similar work can only be successful if the enemy are prevented as far as possible from interfering with the machines engaged on these duties, and such work by hostile machines can only be prevented by interference on our part.

The moral effect of a successful cavalry action is very great; equally so is that of successful fighting in the air. This is due to the fact that in many cases the combat is actually seen from the ground, while the results of successful fighting, even when not visible, are apparent to all. The moral effect produced by an aeroplane is also out of all proportion to the material damage which it can inflict, which in itself is considerable, and the mere presence of a hostile machine above them inspires those on the ground with exaggerated forebodings of what it is capable of doing. On the other hand the moral effect on our own troops of aerial ascendancy is most marked, and the sight of numbers of our machines continually at work over the enemy has as good an effect as the presence of hostile machines above us has bad.

2. SIMILARITY TO FIGHTING ON LAND AND SEA.

To seek out and destroy the enemy's forces must therefore be the guiding principle of our tactics in the air, just as it is on land and at sea. The battle ground must be of our own choosing and not of the enemy's, and victory in the fight, to be complete, must bring other important results in its

(15847.) Wt. W 4610—P.P. 739. 10,000. 4/18. D & S. G. 2. P 18/16⁰ train. These results can only be achieved by gaining and keeping the ascendancy in the air. The more complete the ascendancy, the more far reaching will be the results.

The struggle for superiority takes the form, as in other fighting, of a series of combats, and it is by the moral and material effect of success in such combats that ascendancy over the enemy is gained.

3. Necessity of Offensive Action.

Offensive tactics are essential in aerial fighting for the following reasons:—

    (i) To gain the ascendancy alluded to above.

    (ii) Because the field of action of aeroplanes is over and in rear of the hostile forces, and we must, therefore, attack in order to enable our machines to accomplish their missions, and prevent those of the enemy from accomplishing theirs.

    (iii) Because the aeroplane is essentially a weapon of attack and not of defence. Fighting on land and sea except for the submarine takes place in two dimensions, but in the air we have to reckon with all three. Manœuvring room is, therefore, unlimited, and no number of aeroplanes acting on the defensive will necessarily prevent a determined pilot from reaching his objective. The power enjoyed by the submarine of movement in three dimensions, limited though it is, has to a large extent revolutionized naval warfare.

4. Choice of Objectives.

An aerial offensive is conducted by means of—

    (i) Offensive patrols.

    (ii) The attack with bombs and machine-gun fire of the enemy's troops, transport, billets, railway stations, rolling stock and moving trains, ammunition dumps, &c., on the immediate front in connection with operations on the ground.

    (iii) Similar attacks on centres of military importance at a distance from the battle front or in the enemy's country with a view to inflicting material damage and delay on his production and transport of war material and of lowering the moral of his industrial population.

(i) *Offensive Patrols.*

The sole mission of offensive patrols is to find and defeat the enemy's aeroplanes. Their normal sphere of action extends for some 20 miles behind the hostile battle line, and the further back they can engage the enemy's fighting aeroplanes the more immunity will they secure for our machines doing artillery work, photography and close reconnaissance. Since, however, aerial ascendancy will usually be relative only and seldom absolute, patrols are also required closer in to attack those of his fighting machines which elude the outer patrols, and to deal with his machines doing artillery observation and similar work.

Fighting may take place at any height up to the limit to which the machine can ascend, known as its " ceiling." Artillery observation imposes a limit of some 10,000 feet, but fighting, bombing and photographic machines may fly at any height up to 20,000 feet or even more. Offensive patrols must therefore work echelloned in height (*see* para. 10).

(ii) *Attack of Ground Targets in the Battle Zone with Bombs and Machine-gun Fire.*

The attack of ground targets cannot strictly speaking be described as fighting in the air, but it is an integral part of the aerial offensive designed to weaken the moral of the enemy's troops and cause them material damage. It is carried out by fast single-seater machines flying normally at anything from 100-2,000 feet, either singly or in formation. Fixed targets and, to a certain extent, troops can be attacked with advantage at any time including periods of sedentary warfare, but the attack of moving targets such as troops and transport is of the greatest value in connection with ground operations either offensive or defensive.

(iii) *Attack of Ground Targets at a distance.*

Targets at a distance are usually attacked by bombing. Such raids may be expected to produce their maximum effect when undertaken against distant objectives, since they may cause the enemy to withdraw artillery and aeroplanes from the front for the protection of the locality attacked. They are also, however, of great use in rear of the immediate front in connection with operations on the ground.

Every patrol or raid should, therefore, be sent out with a definite mission, the successful performance of which will not only help us to gain aerial ascendancy by the destruction of hostile aircraft, but will also either tend to induce the enemy to act on the defensive in the air, or further the course of operations on the ground.

4

5. Types of Fighting Machines.

The machines at present in use for offensive purposes may be divided into four main classes:—

    (i) **Fighters,** (*a*) single seater, (*b*) two seaters.
    (ii) **Fighter** reconnaissance.
    (iii) **Bombers.**
    (iv) **Machines** for attacking ground targets from a low altitude.

(i) *Fighters.*

(*a*) Single seaters are fast, easy to manœuvre, good climbers and capable of diving steeply on an adversary from a height.

Their armament consists of two or more machine or Lewis guns, whose axis of fire is directed forward and, usually, in a fixed position in relation to the path of the machine.

Single seater fighters are essentially adapted for offensive action and surprise. In defence they are dependent on their handiness, speed and power of manœuvre. They have no advantage over a hostile single seater as regards armament, and are at a disadvantage in this respect when opposed to a two seater, and, therefore, the moment they cease to attack are in a position of inferiority, and must break off the combat, temporarily at any rate, until they have regained a favourable position. On the other hand, provided they are superior in speed and climb to their adversary, they can attack superior numbers with impunity, since they can break off the combat at will in case of necessity.

(*b*) Two seater fighters have, in addition, a machine gun for the observer, on a mounting designed to give as wide an arc of fire as possible, especially to the flanks and rear. Their front gun or guns remain, however, their principal armament.

The two seater is superior in armament to the single seater, since it is capable of all-round fire, but is generally somewhat inferior in speed, climb and power of manœuvre. It has greater powers of sustaining a prolonged combat, being less vulnerable to attacks from flanks and rear, but as in the case of single seaters its chief strength lies in attack.

When fighting defensively or when surprised in an unfavourable position, it is often best for the pilot to fly his machine in such a way as to enable the observer to make the fullest use of his gun, while awaiting a good opportunity to regain the initiative.

(ii) *Fighter Reconnaissance Machines.*

The first duty of these machines is to gain information.
They do not go out with intent to fight, but must be
capable of doing so, since fighting will often be necessary
to enable the required information to be obtained. Those
at present in use are two seaters, the pilot flying the
machine and the observer carrying out the reconnaissance.
They approximate to the two seater fighter type, and in
the case of missions which can be carried out at 15,000 feet
or upwards, are capable of acting alone, and usually do so.

(iii) *Bombing Machines.*

Bombing machines usually carry at least one passenger, so
that they can, in case of necessity, undertake their own
protection, even when loaded. Their requirements, as
regards armament, are similar to those of fighter reconnaissance machines. Machines carrying more than one
passenger usually have a gunner both fore and aft, and are
strong for defensive fighting. The greater weight of bombs
they can carry the better.

(iv) *Machines for Attacking Ground Targets.*

Machines for this purpose will, as a rule, be single seaters.
Climb is of relatively minor importance, but they require
to be fast and very manœuvrable and must have a very good
view downwards. Single seater fighters can be used for
this work, but it is probable that a special type of machine
will be evolved in which the pilots and some of the most
vulnerable parts will be protected by armour. They will
probably be adapted for carrying a few light bombs.

PRINCIPLES OF AERIAL FIGHTING.

6. FACTORS OF SUCCESS.

The success of offensive tactics in the air depends on
exactly the same factors as on land and sea. The principal
of these are:—

(i) Surprise.
(ii) The power of manœuvre.
(iii) Effective use of weapons.

7. SURPRISE.

Surprise has always been one of the most potent factors of
success in war, and although it might at first sight appear
that surprise is not possible in the air, in reality this is
by no means the case. It must be remembered that the
aeroplane is working in three dimensions, that the pilot's
view must always be more or less obstructed by the wings

and body of his machine, and that consequently it is often
an easy matter for a single machine, or even two or three
machines, to approach unseen, especially if between the
hostile aeroplane and the sun. Fighting by single machines
is, however, rapidly becoming the exception (*see* paragraph 10), and surprise is more difficult of attainment by
machines flying in formation, though by no means impossible.

Even when in view, surprise is possible to a pilot who
is thoroughly at home in the air, and can place his machine
by a steep dive, a sharp turn. or the like, in an unex-
pected position on the enemy's blind side or under his tail.

A surprise attack is much more demoralizing than any
other form of attack and often results in the pilot attacked
diving straight away or putting his machine into such a
position that it forms an almost stationary target for a
few seconds, and thus in either case affords the assailant
an easy shot. To achieve surprise it is necessary to see
the enemy before he sees you. To see other machines in
the air sounds an easy matter. but, in reality, it is very
difficult and necessitates careful training. The ground
observer is guided by the noise of the engine. but the pilot,
of course, hears no engine but his own. Again, while the
ground observer sees the machine, broadly speaking, in plan,
the pilot sees it in elevation, presenting a very much smaller
surface. Add to these the variety of back ground, clear or
cloudy sky or the chequered appearance of the ground from
above, and the obstruction offered to the pilot's view by the
wings and fuselage of his machine and the difficulties will
begin to be realized.

Every pilot must, therefore. be trained to search the sky,
when flying, in a methodical manner. A useful method is
as follows:—Divide the sky into three sectors by means of
the top plane and centre section struts. and sweep each
sector very carefully. From port wing tip to centre
section search straight ahead and then do the same from
centre section to starboard wing tip. From starboard
wing tip take a steady sweep straight upwards to port wing
tip. In addition it is essential to keep a good lookout to
the rear, both above and below the tail, in order to avoid
being surprised. This can be done by swinging from side to
side occasionally. The results of a concentrated search of
this description are surprising. while a pilot who just sweeps
the sky at random will see little or nothing.

In addition to seeing the hostile machine it is necessary
to recognise it as such. A close study of silhouettes will
assist pilots to do this, but until thoroughly experienced it
is a safe rule to treat every machine as hostile. This, of
course, necessitates going close enough to make sure, and

soon results in a pilot becoming familiar with all types of machines in the air.

The types of hostile aeroplanes must be carefully studied, so that the performance and tactics of each, its blind side, and the best way to attack it, can be worked out. Some machines have a machine-gun mounted to fire downwards and backwards through the bottom of the fuselage.

Every advantage must be taken of the natural conditions, such as clouds, sun, and haze, in order to achieve a surprise.

If observed when attempting a surprise it is often best to turn away in the hope of disguising the fact that an attack is meditated. Flat turns may cause the enemy to lose sight of a machine even after he has once spotted it, as they expose much less surface to his view than do ordinary banking turns.

8. POWER OF MANŒUVRE.

Individual skill in manœuvre favours surprise, as pointed out above. Individual and collective power of manœuvre are essential if flying in formation is to be successful or even possible. They can only be obtained by constant practice.

To take full advantage of manœuvre the highest degree of skill in flying and controlling the machine is of the first importance. A pilot who has full confidence in his own powers can put his machine in any position suitable to the need of the moment, well knowing that he can regain control whenever he wishes. The best way to gain the required confidence is for the instructor to take his pupil up, dual control, throw the machine out of control himself and allow the pupil to right it, the instructor only retaking control should the pupil fail to regain it. Once confidence has been acquired practice will make perfect.

The second essential is that the pilot shall know his engine and how to get the best out of it, and thoroughly understand the use of his throttle. Many a chance is lost through pilots allowing their engine to choke in a dive, and no pilot can become really first-class unless he acquires complete practical familiarity with his engine by constant study and practice.

Good formation flying can only be carried out by pilots who know how to use their throttle. The leader must always fly throttled down or his formation will straggle while they in their turn must make constant use of their throttle to maintain station and twist, turn and wheel without confusion or loss of distance.

Other points to which attention must be paid are the following:—Pilots must know the fuel capacity of their

machine and its speed at all heights. The best height at
which to fight varies with each type of aeroplane. Each
pilot must know this height so that he can make the very best
use of his machine. As a general rule machines should patrol
at a greater altitude than their best fighting height. The
direction and strength of the wind must be studied before
leaving the ground and during flight. This study is most
important since wind limits the range of action, as machines
when fighting are bound to drift down wind.

Knowledge of the ground and ability to read a map and
use a compass are of extreme importance. When engaged
in fighting it is impossible to watch the ground, and unless
pilots acquire an eye for country by constant practice and
thoroughly understand map reading and the use of the
compass they will always have difficulty in picking up their
bearings after a combat. The sun when visible is a valuable
guide to direction.

9. Effective Use of Weapons.

(i) Machine and Lewis Guns.

The essentials for successful fighting in the air are skill
in handling the machine and a high degree of proficiency
in the use of the gun and sights. Of these two essentials
the second is of even more importance than the first. Many
pilots who have not been exceptionally brilliant trick fliers,
have had the greatest success as fighting pilots owing to
their skill in the use of the gun and sights. The manipula-
tion of a gun in the air, especially on single-seater machines,
is a very much more difficult matter than on the ground.
Changing drums, for instance, though simple on the ground,
is by no means easy when flying.

Every pilot and observer who is called upon to use a
machine gun must have such an intimate knowledge of its
mechanism as to know instinctively what is wrong when a
stoppage occurs, and, as far as the type of machine allows,
must be able to rectify defects while flying. This demands
constant study and practice both on the ground and in the
air.

It is absolutely essential that pilots and observers should
know exactly how their guns are shooting, and they should
be tried on a target at least once a day. With his gun out
of action a pilot or observer is helpless either for offence
or defence.

Aerial gunnery is complicated by the fact that both gun
and target are moving at variable speeds and on variable
courses. Consequently, however skilful the firer, he cannot

hope to be dead on the target for more than a very few seconds at a time, and it is essential that hand, eye, and brain be trained to work together.

Accurate shooting on the ground from a fixed gun at a fixed target is the first step in training; subsequently constant practice on the ground both when stationary and when moving at fixed and moving targets is essential. Finally, every opportunity must be taken of practice in the air under the conditions of a combat.

Except at point blank range, it is essential to use the sights if accurate fire is to be obtained, and constant practice is needed with the sights provided. The aim can be checked with absolute accuracy by means of the gun camera, and combats in the air during which the camera is used are a most valuable form of training.

Tracer ammunition is of some assistance, but must be used in conjunction with the sights, and not in place of them. Not more than one bullet in three should be a tracer, otherwise the trace tends to become obscured. Too much reliance must not be placed on tracer ammunition at anything beyond short range. The principle should be to use the sights whenever possible at all ranges.

Inexperienced pilots are too apt to be content with diving and pointing their machine at the target and ignoring everything else. Mere noise and fright will not bring down an opponent; it is necessary to hit him in a vital spot. From the time a pilot starts to dive he should not have to fumble about for triggers and sights. His eye should fall automatically on the sight and his hand close on the trigger. By holding the right arm firmly against the body and working only from the elbow the machine can be held much steadier in a dive.

(ii) *Bombs.*

Skill and accuracy in bombing in the same way can only be acquired by continual practice and careful study of the conditions which govern the correct setting and use of bomb sights. Such practice is best obtained by the use of the Batchelor Mirror or of the camera obscura, and must be carried out from varying altitudes up to 15,000 feet, from which height bombs will often have to be dropped on service.

An exception must be made in the case of bombing by single-seater fighting machines from a low altitude, a method of attack which has been employed with very considerable success. In this case no sight is used, and the method found by experience to give the best results is to dive the machine steeply at a point on the ground a few yards in front of the target. The lag of a bomb released from a few hundred feet on a steep dive is very little. Individual pilots must find out by experiment exactly how far ahead they must aim.

FORMATION FLYING.

10. EVOLUTION OF FORMATION FLYING.

The development of aerial fighting has shown that certain fundamental maxims which govern fighting on land and sea are equally applicable in the air. Among these are concentration and mutual co-operation and support. The adoption of formation flying has followed as an inevitable result.

Any mission which has fighting for its object, or for the accomplishment of which fighting may normally be expected, must usually, therefore, be carried out by a number of machines, the number depending on the amount of opposition likely to be encountered and on a third fundamental axiom, namely, that no individual should have more than a limited number of units under his immediate control.

The evolution of formation flying has been gradual. When aerial fighting became general it was soon discovered that two machines working together had a better chance of bringing a combat to a decisive conclusion than had a single machine. The next step was for two or more pairs to work together and this quickly became the accepted practice.

The chief difficulty is control of the remaining machines by the leader primarily due to the difficulty of communication in the air. For practical purposes this limits the number of machines that can be controlled by one man to six, and even when wireless telephony between machines is perfected this number is unlikely to be exceeded. The principles and causes which have led to formation flying remain in force, however, and are bound to result in a further development in the case of offensive fighting, namely, two or more formations working in close co-operation with each other and the best means of achieving such co-operation is the next problem to be solved in aerial warfare.

When a force on the ground is engaged in offensive action the troops comprising the main body must be protected from surprise from the front, flanks and rear. Hence the universal employment in open warfare of advanced flank and rear guards. In the air the third dimension renders flank and rear guard unnecessary, their place being taken by the " Above Guard," which can perform the duties of both. Whether we consider a single formation, therefore, or a group of formations acting in close co-operation, an " Above

Guard " is necessary and may consist of two or more machines in the first case or of one of the formations in the second. These should fly slightly above the main body, either directly behind or echelonned to a flank. The main body carries out the offensive fighting, the " Above Guard " remaining intact above them to protect them from surprise.

11. SOME PRINCIPLES OF FORMATION FLYING.

The formations adopted vary in accordance with the mission and with the type of machine. Certain principles are, however, common to all formation flying and must be strictly observed.

(i) As on the ground so in the air the bed-rock of successful co-operation is drill, and good aerial drill is an essential preliminary to success in formation flying for any purpose. Before commencing drill in the air it has been found of great assistance to practice on the ground until all concerned are thoroughly conversant with the various evolutions. Simplicity is essential and complicated manœuvres are bound to fail in action. Drill should commence in flight formation, each Flight Commander instructing and leading his own flight. Subsequently the Squadron Commander should lead and drill his whole squadron in three flights, each under its Flight Commander. A really well-drilled flight can manœuvre in the air with as little as a span and a half between wing tips, but in action it is better to keep a distance of 80 to a 100 feet, otherwise pilots are apt to devote too much of their attention to avoiding each other.

(ii) One of the first essentials of successful formation flying is that every pilot thoroughly understands the use of his throttle. He will have to use it constantly throughout the flight and must train himself to do so instinctively. The throttle must be used to keep station. If a pilot attempts to do so by sharp turns instead of by using his throttle he will inevitably throw the formation into disorder.

(iii) The formation adopted must admit of quick and easy manœuvre by the formation as a whole.

(iv) A leader must be appointed, and a sub-leader, in case the leader has to leave the formation for any reason, e.g., engine trouble. The machines of leaders and sub-leaders must be clearly marked. Streamers attached to different parts of the machine are suitable. Good formation flying depends very largely on the leader, who must realise that his responsibilities do not end with placing himself in front for others to follow. Their ability to do so depends very largely on himself and on constant practice together so that they know intuitively what he will do in any given circumstances.

(v) An air rendezvous must be appointed, and the leader must see pilots and observers before leaving the ground and explain his intentions to them. To save waste of time in picking up formation in the air and to ensure a really close formation, machines must leave the ground together or as nearly so as possible and in approximate formation. When all machines have reached the rendezvous, the leader fires a signal light, indicating that formation is to be picked up at once. He should then fly straight for a short time, as slowly as possible, while his observer, if he has one, reports on the formation. If one or more machines are rather far behind, the leader should turn to the right or left, after he or his observer has given a signal that he is going to do so. Thus the machine behind will be enabled to cut a corner and close up. When the leader is satisfied with the formation he fires a light signifying that he is ready to start. The actual signal to start can be given either by the leader or from the ground; in the latter case the officer on the ground, who is responsible for the despatch of the formation, will also be responsible for deciding when the proper formation has been adopted. The decision as to the suitability or otherwise of the weather conditions will in any case rest with the leader of the formation. A suitable code of signals for formation flying is given in Appendix A. Signal lights must be fired upwards by the leader, otherwise machines in the rear may have difficulty in seeing them.

(vi) Pilots must clearly understand how the formation is to reform after a fight. Once an attack has been launched, it must tend to become a series of individual combats, but if a formation is able to rally at the first lull and make a second concerted attack, it should gain a real advantage over a dispersed enemy formation. Definite instructions by the leader on the point are essential. A rendezvous over a prearranged spot has been found suitable, in the case of a small area. In the case of a large area two or more spots may be designated previously, the rendezvous to take place over the nearest. It must be realised that pre-arrangements may be found unsuitable, and in every case each pilot must invariably close on the nearest machine. If there is a choice he will join two machines in preference to a single machine and three machines in preference to two. This applies to the leader also. To rendezvous successfully after a fight needs continual practice.

(vii) Formations must not open out under anti-aircraft gun fire. It has been found by experience that fire is usually less effective against a well closed up group of machines than when directed on a single machine. To open out is to give the enemy the chance, for which he is waiting, of attacking the machines of the formation singly. The enemy's aim can be thrown out temporarily, if the fire is very hot, by turning sharply, diving or climbing, but it is seldom advisable to lose height, especially when far over the enemy's lines.

(viii) The formation should be retained until the aerodrome is reached on the return journey.

12. USE OF FORMATION FLYING.

Flying in formation is necessary in the case of :—

(i) Offensive patrols;
(ii) Bomb raids;

and is the normal method of carrying out these duties.

Medium and long distance reconnaissances may also have to be carried out in formation, but a fast machine capable of flying at a great altitude can often carry out such reconnaissances by itself, including photography when large scale photographs are not required. A further development of formation flying is in the attack of ground targets with machine-gun fire (*see* para. 15).

13. OFFENSIVE PATROLS.

The sole duty of offensive patrols is to drive down and destroy hostile aeroplanes, and they should not be given other missions to perform, such as reconnaissances, which will restrict their fighting activities. In the face of opposition of any strength offensive patrols usually have to fly in formation in order to obtain the advantage of mutual support, but the formations adopted can be governed solely by the requirements of offensive fighting. Single seater scouts or even two seaters, if superior in speed and climb to the great majority of the enemy's machines, may be able to patrol very successfully alone or in pairs, taking advantage of their power of manœuvre and acting largely by surprise, but in the case of machines which do not enjoy any marked superiority formation flying is essential. Fighting in the air, however, even when many machines are involved on each side, tends to resolve itself into a number of more or less independent combats, and it has been found advisable to organize a purely fighting formation accordingly. Such a formation can suitably consist of six machines, organized in groups of two or three machines each, every group having its own sub-leader, the senior of whom takes command of the formation. A deputy leader should also be designated, in case the leader falls out for any reason. As far as possible the groups should be permanent organizations, in order that the pilots may acquire that mutual confidence and knowledge of each other's tactics and methods which is essential for successful fighting. It must be impressed on pilots that the group is the fighting unit and not the individual (see paragraph 17).

14. Reconnaissances and Bomb Raids.

In reconnaissance the whole object is to protect the reconnaissance machine or machines, and enable them to complete their work. Opposition will usually take one of two forms. The enemy's scouts may employ guerilla tactics, hanging on the flanks and rear of the formation, ready to cut off stragglers, or attacking from several directions simultaneously; or else the formation may be attacked by a hostile formation. The modern type of two seater fighter reconnaissance machine is able to deal with either class of opposition without assistance. The machines must fly in close formation, keep off enemy scouts which employ guerilla tactics by long range fire, and be ready to attack a hostile formation if the enemy's opposition takes that form.

By skilful manœuvring it may be possible to bring a superior number of guns to bear on a portion of the enemy's formation than he can bring into action owing to the fire of some of his aeroplanes being masked by the machines in front of them.

Reconnaissance formations, like fighting formations, can be organized in groups, each with its sub-leader, but as the object is to secure the safety of the reconnaissance machine, the whole formation must keep together and act as one.

A suitable formation in the case of six two seater machines has been found to be two lines of three, the flankers in the front line being slightly higher than the centre (reconnaissance) machine, and the three machines in rear slightly higher again. The intervals between the machines should not be more than 100 yards, and the distance of the rear rank from the front should be sufficient only to admit of a good view being obtained of the leading machines.

The pace must be slow, otherwise the rear machines are bound to straggle. Machines must, therefore, fly throttled down. Sharp turns by the leader also lead to straggling; a signal, therefore, should always be given before turning, and a minute or two allowed, if possible, after giving the signal before the turn is commenced, in order to give the machines on the outer flank time to gain ground.

The duty of bombing machines is to get to their objective and to drop their bombs on it, and only to fight in the execution of their duty. The secret of success is the most careful pre-arrangement, so that everyone knows exactly what he has to do. The bombing machines, like a reconnaissance, must keep in close formation. Any tendency to straggle or to open out under anti-aircraft fire will give the enemy the opportunity he is seeking to attack and split up the formation. A well-kept formation, on the other hand, is seldom attacked at close range, unless by very superior numbers. When bombing from a height the best results have been secured by dropping bombs while still in formation. Three machines drop their bombs simultaneously, the centre observer being responsible for the sighting or, if preferred, all machines can drop their bombs simultaneously on a signal from the leader. If it is necessary for machines to break formation to drop their bombs, a rallying point must always be chosen beforehand where they will collect and resume flying formation as soon as their bombs have been released.

When a very large raid is contemplated, it will often be best to carry out the attack by two separate formations, since there is a limit to the number of machines which can be controlled efficiently by a single leader. Six bombing machines are normally the maximum. The departures of the two formations from their respective rendezvous, if they are to make a single raid, should be so arranged as to enable them to give one another mutual support in case of a heavy hostile attack. The rendezvous should not be too close together, 10 to 15 miles apart is a suitable distance. Departures from the rendezvous should be timed so that the first formation is leaving the objective as the second approaches, and the leaders should watch each other's signals.

With modern machines an escort to a reconnaissance formation or bomb raid is seldom desirable, and far better results are obtained by sending one or more offensive patrols to work independently over the area where opposition to the reconnaissance or raid is most likely to be encountered. If an escort is provided, its primary duty is to enable the reconnaissance or raid to accomplish its mission and it should only fight in the execution of this duty. It is usually best to keep the escort and the machines it is protecting as distinct formations under a separate leader. The escort flies above the reconnaissance or bombing machines, in such a position as to obtain the best view of them and the greatest freedom of manœuvre in any direction. Its rôle is:—

   (i) To break up an opposing formation.

   (ii) To prevent the concentration of superior force on any part of the formation they are protecting.

   (iii) To assist any machine which drops out of the formation through engine or other trouble.

While the bombs are being dropped, the escort should circle round above the bombing machines, protecting them from attack from above, and ready to dive on to any hostile machine that may interfere with them.

15. ATTACK OF GROUND TARGETS.

Formation flying has lately been adopted for the attack of ground targets with excellent results, formations appearing to be no more vulnerable to rifle and machine-gun fire from the ground than is a single machine. This is probably due to a tendency to fire at the formation as a whole instead of picking out a particular machine. On the other hand, a formation, as against a single machine, possesses the following advantages:—

   (a) There is less chance of machines losing their way as there are several individuals instead of one only attempting to keep their bearings.

   (b) A greater volume of fire is brought to bear on any target discovered.

   (c) A formation is stronger if attacked.

   (d) A formation may be expected to have greater moral effect on the enemy's troops.

Formation flying at low altitudes demands even more constant practice together than does formation flying at a height, because fire from the ground makes continuous changes of direction and height a necessity. A suitable height from which to attack ground targets is 600 to 800 feet. The essential point is to go low enough to make certain of differentiating between our own and the enemy's troops. Above 800 feet this is difficult, and the chance of interference by hostile aircraft is greater, but these seldom come down to fight below 1,000 feet. Formations for low flying should never exceed six machines.

FIGHTING TACTICS.

16. GENERAL.

Fighting tactics vary with the type of machine and with the powers and favourite methods of individual pilots. No hard-and-fast rules can be laid down, but the following hints based on the experience of others may be of use to the young pilot until he has acquired experience of his own. There are four golden rules which are applicable to all offensive aerial fighting:—

(a) Every attack must be made with determination and with but one object, the destruction of the opponent.

(b) Surprise must be employed whenever possible.

(c) **If surprised or forced into an unfavourable position a pilot must never, under any circumstances, dive straight away from his opponent.** To do so is to court disaster, since a diving machine is an almost stationary target. Moreover, the tactical advantage of height is lost by diving and the initiative surrendered to the hostile machine. The best course of action depends on the type of machine and is discussed below.

(d) Height invariably confers the tactical advantage.

17. SINGLE SEATER FIGHTING.

Fighting in formation with single seaters is a most difficult operation and demands constant study and practice, the highest degree of skill on the part of the individual pilots, mutual confidence between them, and intimate knowledge of each other's methods.

The patrol leader's work consists more in paying attention to the main points affecting the fight than in doing a large share of the fighting himself. These main points are:—

(i) The arrival of more hostile machines, which have tactical advantage, i.e., height.

(ii) The danger of the patrol being carried by the wind beyond the range of its petrol supply.

(iii) The patrol getting below the bulk of the hostile formation.

As soon as any of these conditions occur it is usually better to break off the fight temporarily, and to rally and climb above the enemy before attacking them again.

When fighting in formations of two or more groups, the fighting unit should be the group, each selecting its own objective and acting as described below. The groups will often become separated, but every effort should be made to retain cohesion within the groups. The practice of individual pilots breaking away from the formation to attack hostile machines almost always leads to disaster sooner or later. If the enemy machines scatter, attention should be concentrated on those lagging behind, and, if they dive and are followed down, at least one group should remain at a height as a protection from surprise.

The dangerous quarter in the case of a formation of single seaters is the rear, and care must always be taken to keep a constant watch behind and above. If surprised in an unfavourable position it should be the invariable rule, if time permits, to turn and attack the adversary before he comes to close quarters. If, however, he succeeds in doing so, the best chance lies in a quick climbing turn. Any method which entails losing height such as a side-slip or a spin is bad, as the hostile machine has merely to follow and attack afresh from above.

Surprise by a formation is difficult, and success must be sought in close co-operation and boldness of attack. If the enemy is inferior in numbers an opportunity will occur for a concerted attack by a group against a single machine. If working in groups of three the actual attack should be carried out by two machines, the third remaining above to protect them from surprise. The two attacking machines may converge on the enemy from different directions on the same level, but the attacks must be simultaneous so that they cannot be engaged separately. Another method is to attack echelonned in height, the lower machine diving and attacking the enemy from behind, while the upper machine awaits an opportunity to swoop down on him when he turns to engage the machine that attacked first.

An attack of equal numbers will usually resolve itself into a series of individual duels. The leader must always ensure that his formation is well closed up before attacking, giving the rear machines time if necessary, so that all pilots can attack their adversaries simultaneously.

In attacking superior numbers the best chance of success lies in the destruction of the enemy's moral by excessive boldness.

Decoy tactics are sometimes successful. One group attempts to draw the enemy on to attack, while the other flies high above it, ready to surprise the enemy should he seize the apparent opportunity. Watch must be kept for similar tactics on the part of the enemy.

The group going down as a decoy must not be more than about 3,000 ft. below the remainder or it will run the risk of being attacked from the flank by superior numbers before the groups above can get down to its assistance.

If, owing to being cut off from his formation and being attacked by a superior number of machines, a pilot is forced down low, his best method of escape is usually to go down quite close to the ground and fly back on a zigzag course.

Although as a principle single-seaters should not act alone, yet in many cases isolated scouts will be called upon to fight single-handed, e.g., when a formation has become split up during a combat and a machine fails to rejoin its formation. Again, selected pilots on the fastest types of single-seaters may be usefully employed on a roving commission, which will enable them to make the greatest use of surprise tactics.

Single-seater fighting calls for much initiative, especially when a combat develops itself into individual fighting and the pilot has the opportunity of developing his own particular method of attack. Methods vary with the type of machine attacked, and may be conveniently discussed under two headings:—

(a) Single-seater against single-seater.

(b) Single-seater against machines with one or more passengers.

(a) Single-seaters are best attacked from above and behind with a view to getting within point blank range if not observed. Height enables the attacker to anticipate his enemy's movements more quickly and to guard himself from attack from behind by a sudden turn on the part of his opponent. It is therefore essential to have plenty of engine power in hand so as to keep the means of climbing above the enemy throughout the fight and thus retain the advantage of height whatever tactics he may pursue. When attacking a hostile formation, one of their number, more often than not their leader, will sometimes fly out of the fight and climb his utmost with a view to getting above the attackers. The leader of the attacking formation should watch for this manœuvre, and be ready to frustrate it by climbing himself.

The knowledge that there is one enemy above not only nullifies the advantage in height but divides the attention of the attacking pilots just when it should be entirely concentrated on the machines they have severally selected to attack.

A hostile pilot who attempts to come up unawares from behind and below can usually be defeated by a quick climbing turn. He will often be taken by surprise and turn flat, offering a vulnerable target to attack from above.

Attacks from directly in front or from the flanks are often successful, as the vital parts of the machine from the pilot forward are fully exposed. Aim should be taken at the front of the machine in such an attack. It is a common mistake to aim at the pilot, which usually only results in hitting the fuselage, as the majority of the fire usually takes effect behind the point of aim. This is conclusively proved by the number of our machines which return with the fuselage riddled and little or no damage from the pilot forward.

Similarly, when attacking from above and behind, aim should be taken at the leading edge of the top plane, thus increasing the chance of hitting the engine and pilot.

When it is necessary to swerve to avoid a collision or to break off the combat temporarily to change a drum or rectify a jamb, this should be done by a sudden turn or climb, care being taken subsequently to avoid flying straight or losing height. When ready a favourable position must be regained by manœuvre before renewing the attack.

(b) Single seaters attacking two-seaters can do so from behind and above, from behind and below, or from front or flanks. The most favourable method is perhaps to attack from behind and below attempting to achieve surprise by climbing up under the fuselage and tail plane, the blindest spot from the point of view of the observer. A skilfully handled single-seater which can attain a position about 100 yards behind and 50 feet below a hostile two-seater without being observed, is in a position to do most damage to the enemy with least risk to himself. Once in this position the object of the attacker must be to keep out of the enemy's field of fire as much as possible. The two-seater will endeavour to bring fire to bear on the attacker by turning quickly in order to deprive him of the cover of the fuselage and great skill is required to retain a position directly in rear in spite of frequent turns. If enjoying superior speed, which will usually be the case, the single-seater should turn always in the opposite direction to the two-seater, *e.g.*, if the two-seater turns to the right, the

attacker at once turns to the left, thus preserving their
relative positions. When on the bank in the act of turning,
the two-seater will offer a favourable target to the attacker
if the latter is quick enough to take advantage. A short
quick burst at this moment may confuse the pilot and cause
him to dive, in which position it will be very difficult for
the observer to do any accurate shooting, or even to stand
up to fire, owing to the wind pressure, and it is safe to
disregard the rear gun for the time being. Should the
observer be put out of action the rear gun can, of course,
be disregarded altogether and the attacker can close to
point-blank range.

When attacking two-seaters from above a short steep dive
is effective, because the gunner has then to shoot almost
vertically upwards, which is difficult and impairs the accuracy
of his aim. To dive from behind and above otherwise than
steeply, on the other hand, is to afford the hostile observer
a particularly easy shot. If approaching head on with a
view to turning and attacking from behind, the turn must
be made before a position vertically over the opponent is
reached, otherwise the attacker will be left two or three
hundred yards behind the hostile machine with no chance of
surprise and in not a very favourable position for attack.
An attack from the front and above or from the flanks pre-
cludes the use of the observer's gun altogether in many
types of machines, but care must be taken not to give the
observer an easy shot by diving straight on past the
machine after delivering the attack.

Surprise can often be attained by carefully watching the
adversary, preferably from behind. An especially favourable opportunity for surprise occurs in the case of a hostile
machine crossing our front on some special mission, for
once the hostile observer has satisfied himself that the air
is clear, he will give his principal attention to his work.
The enemy will often choose cloudy weather for such missions,
and this gives special chances of surprise to a skilful pilot,
working with intelligence. In such weather it must be
remembered that it is often of advantage to approach the
hostile machine on his own level when the planes form but
a thin line which is difficult to see. When surprise is impossible, advantage must be taken of the handiness and
manœuvring power of the scout to prevent the enemy from
taking careful aim by approaching him in a zigzag course,
and never in a straight line, since a machine attacking in
a straight line offers a comparatively easy target. When
within about 100 yards the zigzag course must be abandoned,
and the moment when the enemy is in the act of shifting his

aim should, if possible, be chosen. He can then be attacked in a straight line with a burst of rapid fire, or it may be possible to get below him and fire at him more or less vertically at almost point-blank range.

To open fire at long range is to give the advantage to the enemy, since it is necessary to fly straight to bring fire to bear, and an easy mark is thus offered.

In the case of a group of three attacking a single two-seater, as in that of single-seaters, one machine must remain as an "Above Guard." The other two will have a very good chance of surprise if one machine repeatedly makes short dives firing a few rounds and climbing again. This will engage the attention of the observer and afford the second machine an opportunity of creeping up underneath the enemy to point blank range.

In the attack of multi-seater machines, surprise is even more essential to success, since they usually have a gun on a circular mounting both in front and rear, and consequently have practically no blind spot. Some types have also a gun mounted to fire downwards at an angle through the fuselage in order to deal with attack from behind and below.

18. Two-Seater Fighting.

The principles of fighting in two-seaters designed for the purpose are similar to the above, but in the actual combat they are able to rely more on their power of all round fire and less on quickness of manœuvre. The fighting tactics adopted should, therefore, be such as to favour the development of fire. The single-seater, when no longer able to approach its adversary, temporarily loses all power of offence and has to manœuvre to regain a favourable position. The two-seater, on the other hand, can develop fire from its rear gun, after passing its adversary or on the turn. The gun or guns firing straight ahead must be looked on as the principal weapons, the fire of the observer being brought to bear after passing the adversary, on a turn or against another machine attacking him from the rear.

A two-seater like a single-seater must, however, never dive straight away from an adversary, as even though it can fire to the rear the advantage is all with the machine which is following.

Formations of two-seaters are less liable to surprise from the rear, since the observers of the rear machines can face in that direction and keep a constant look-out. Mutual fire support is also easier in their case, in view of their all-round fire. They are, therefore, as already pointed out, better able to sustain a protracted battle.

The essence of successful fighting in two-seaters lies in the closest co-operation between pilot and observer. They must study their fighting tactics together, and each must know what the other will do in every possible situation.

The tactics of an artillery or bombing machine should be more defensive in their nature since their primary work is not to fight but to fulfil their mission. Machines of these types are also usually at a considerable disadvantage as regards quickness of manœuvre. They should therefore be fought in such a way as to give the observer every chance of bringing effective fire to bear, and the front gun should be retained for use when opportunity offers, such as when a hostile machine, attacking from behind, overshoots the mark.

19. Fire Tactics.

Opportunities in the air are almost invariably fleeting, and consequently the most must be made of them when they occur. Fire should therefore, be reserved until a really favourable target is presented, and should then be in rapid bursts. Fire should only be opened at ranges over 300 yards when the object is to prevent hostile machines coming to close quarters, as in the case of an escort to a reconnaissance machine, and should not be opened at ranges over 500 yards under any circumstances. In offensive fighting the longer fire can be reserved and the shorter the range, the greater the probability of decisive result.

For an observer on a two-seater machine, however, a range of from 200 to 300 yards is suitable, since it enables full advantage to be taken of the sight. Fire may be opened at longer range when meeting a hostile machine than when overhauling it, otherwise there will be no time to get in more than a very few rounds owing to the speed with which the machines are approaching one another. Pilots and observers must accustom themselves to judging the range by the apparent size of the hostile aeroplane and the clearness with which its detail can be seen. This needs constant practice.

A reserve of ammunition should be kept for the return journey when fighting far over the lines.

Manœuvre is an integral part of fire tactics, and every endeavour must be made to manœuvre in such a way as to create favourable opportunities for one's own fire and deny such opportunities to the enemy.

APPENDICES

APPENDIX A.

CODE OF LIGHT SIGNALS TO BE USED IN FORMATION FLYING.

Colour.	Fired by.	Indicates.
Red ...	Leader, in conjunction with K strips and red light from ground.	Leave the rendezvous. Leader fires a light to indicate he is ready to leave the rendezvous. The formation leaves on this signal or awaits an order from the ground consisting of a K and a red light.
White ...	Leader, or from the ground in conjunction with N in strips.	Return to your aerodrome. Expedition abandoned. This signal applies east or west of the line. If fired east of the line it also indicates "Keep formation till line is crossed."
Red ...	Any member of formation.	"I am being attacked and need assisatnce."
Red ...	Leader	Rally to continue operation—(attack having been dispersed).
Green ...	Any member of expedition (including leader).	"I am forced to return to my aerodrome." This signal, if fired by the leader, does not imply that the expedition is abandoned. The leadership must be taken up by the deputy leader.

www.ingramcontent.com/pod-product-compliance
Lightning Source LLC
Chambersburg PA
CBHW020953030426
42339CB00004B/79